More Twig Mosaic Furniture

L.J. HAWKINS

4880 Lower Valley Road, Atglen, PA 19310

Copyright © 1998 by Larry Hawkins
Library of Congress Catalog Card Number: 97-81407

All rights reserved. No part of this work may be reproduced or used in any form or by any means—graphic, electronic, or mechanical, including photocopying or information storage and retrieval systems—without written permission from the copyright holder.

This book is meant only for personal home use and recreation. It is not intended for commercial applications or manufacturing purposes.

Designed by Laurie A. Smucker
Typeset in Times New Roman

ISBN: 0-7643-0499-2
Printed in China

Published by Schiffer Publishing Ltd.
4880 Lower Valley Road
Atglen, PA 19310
Phone: (610) 593-1777; Fax: (610) 593-2002
E-mail: schifferbk@aol.com
Please write for a free catalog.
This book may be purchased from the publisher.
Please include $3.95 for shipping.
Try your bookstore first.

We are interested in hearing from authors with book ideas on related subjects.

Introduction

I love mosaic twig furniture. The textures, the play of light on the bark, the infinite variety of colors...all of them combine to create a truly enchanting piece of functional art. Twig mosaic work is never boring. Every piece is unique, and every twig contributes its own personality to this uniqueness. Together they give the furniture a liveliness and interest that make it a valued addition to the home.

The heart of mosaic twig furniture in America was the Adirondack region of New York. The lakes and mountains were dotted with homes and lodges used for vacations and other getaways by prosperous urbanites. The craftsmen who built them were local and used local materials...the bounties of the surrounding forests. When the building season ended, these same craftsmen made much of the furniture for the lodges. With the ingenuous use of twigs and bark, they formed primitive yet elegant tables, chairs, beds, bureaus, sideboards and more.

Though its highest point of fashion was the late 19th century, rustic furniture is still treasured by collectors and designers around the world. It is showing up in design centers around the country and is gracing some of the most wonderful homes of the nation. In both its antique forms and in the creations of contemporary artisans, rustic furniture is being appreciated as a true folk art.

Of the rustic forms, the one I love is mosaic twig work. It has an innate beauty and comes in an endless variety. This variety is also there for the craftsperson. No two pieces are alike, just as no two trees are alike. The other great advantage for the builder is that the materials are so readily available. I simply walk out my back door, and they are there. With simple tools it is possible to create the most amazing pieces of furniture. All that remains is the ability to see the possibilities and to carry out a vision.

I hope you find this an art that is enjoyable and rewarding. I know I have. Perhaps this book will be a help in getting you started.

Materials

The case for this chest is a combination of cabinet grade plywood and 4/4 walnut. This gives the piece a solidity and weight that feels right. You could use solid walnut or another hardwood if you desire, and for many of my pieces that is what I choose to do.

I harvest twigs every few days. Because insects quickly attack dead wood, I use only live trees and branches. To avoid warping I let the wood age for 6 months before using it. I store it in a dry garage, standing it upright in bundles. I frankly do not know exactly what trees I use, picking them for color and texture rather than for their pedigree. I do, however, avoid birch, which tends to disintegrate over time. The birch bark, however, is quite durable and beautiful as an accent, as you will see in the drawer treatment on this piece. Birch bark is traditionally used to fill spaces in twig work. In a future volume I will share some wonderful techniques for decorative bark work.

The twigs are nailed to the wood with brads. They need to be long enough to secure the twig. Because I do so many, I use a compressed air gun to drive my brads. A hammer and nailset will do as well, though it will be slower.

The finish should bring out the texture and color of the twigs, not conceal them. I often use tung oil or, as in this project, satin finish polyurethane.

The Project

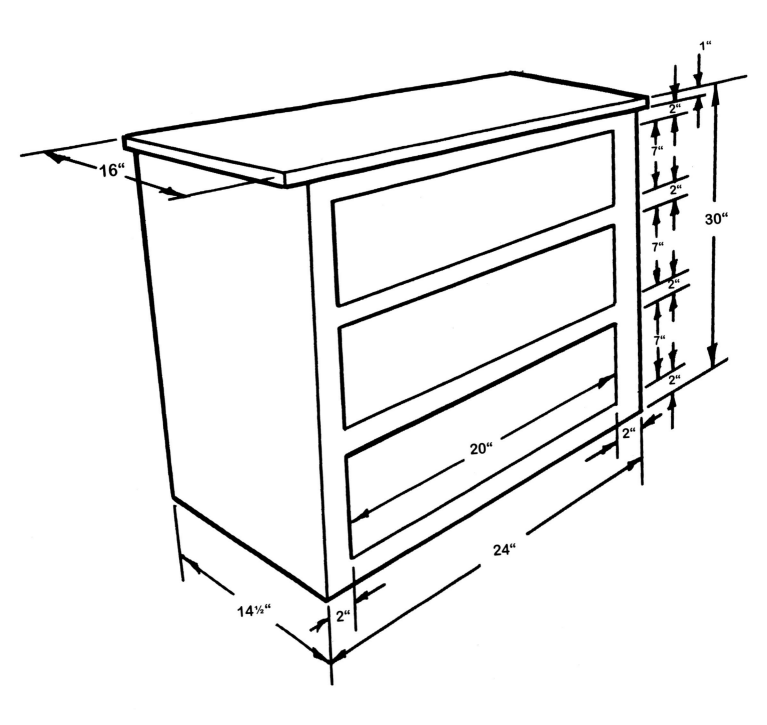

The project is a three drawer chest. The sides and back are of plywood. The front framing and drawers are of 4/4 walnut, as is the top. Overall dimensions of the case are 30" x 24" x 14-1/2". The top measures 26" x 16" x 1". The drawers are 7" and 20" wide.

Measure from the bottom of the piece to the bottom of the drawer.

The view from the side.

Carry the mark around to the side, front...

Find the center from the top to the scribed line and mark at the back edge...

and back...

and the front...

and scribe a line.

Then scribe a line.

Do the same for the vertical center. Find the center front to back.

and cut it a little longer than the measure.

and scribe a line.

Split the twig on the bandsaw. We will save one half for the other side of the chest, so it will match perfectly.

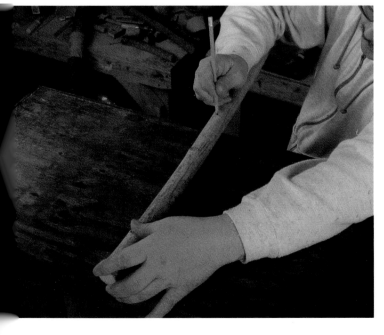

Take a 1" twig, mark the width of the cabinet on it...

Sand the surface flat...

and flatten the edges. The flat edges help achieve a snug fit with adjoining twigs. Ideally there should be no discernable gap between the twigs.

Realign the twig so the bottom edge again is on the center line and use 1" brads to tack the ends in place first. Then continue to nail across the twig, about every 4".

Flatten one end.

Measure another piece of the same diameter for use at the top of the side. As before cut it a little long and split it down the middle, sand the piece, set the final length, cut it...

With the top edge of the twig on the scribed line at the bottom of the chest, hold the sanded end of the twig in place and mark the other end for cutting. Cut it to length.

and nail in place at the top.

A similar piece goes across the middle. Measure carefully from the middle up...

Before nailing the other end of the cross piece, measure up...

and down to make sure it is centered.

and down and adjust for center.

When one end is centered take it in place with a single brad.

To continue framing the sides I measure a piece to fit between the top and middle cross pieces. I will cut it a little long for now and sand it as before.

Do a final fitting, trimming it to the proper length.

The framing complete.

Nail it in place.

Next I'll add a vertical piece in the center of each panel. This is slightly smaller in diameter than the framing pieces. Measure the length.

Do the same on the other edge of the top panel and at the edges of the bottom.

Cut, sand and put into position on the center line. Double check with a measure at the top...

checking each side to make sure they are equal.

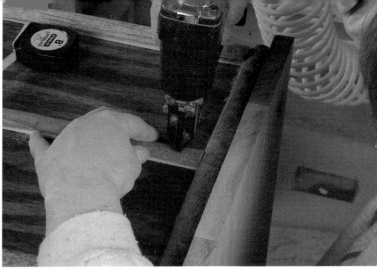

Nail it in place.

tack one end of the twig in place...

When it is correctly positioned, nail it about every two inches.

and center the other end.

Repeat the process in the bottom panel for this result.

Having created four panels, I now want a cross piece at the top and bottom of each one. Return to the large diameter of the framing twigs, and cut and sand them to size. Nail one in place...

Mark 4 pieces to fit between the two twigs you just added to the panels.

at each end of the panel.

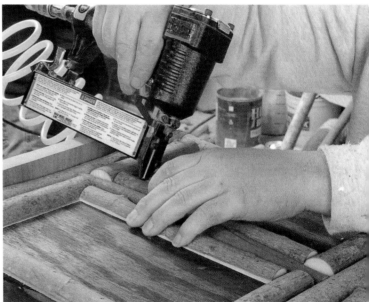

Nail them in place next to the center piece in each panel. To draw the twigs close it is helpful to nail at an angle toward the piece that is already in place.

Progress.

The result.

Progress.

Measure 8 pieces to fit from the center out to the side twig, leaving them long for now. Cut, split, and sand them.

Again, measure for twigs that will fit between the top and bottom pieces in each panel.

Make a final fit and nail one across the top and bottom of each panel.

Cut, sand, fit and apply along the inside vertical twigs. A rubber mallet is useful in achieving a tight fit.

Progress.

and vertical pieces along the center.

Occasionally I like to leave twig knobs and other imperfections in the wood to give the piece character.

As you get to the end, use smaller diameter pieces.

Continue the pattern. Horizontal pieces top and bottom...

This gives you room for one more vertical piece.

Complete the diamond pattern with the last vertical twig...

Moving to the front of the chest, we need to break the horizontal members. I'll do this with some small vertical pieces. First, though, I will run a horizontal piece under the bottom drawer. Mark the center.

and the last two horizontal twigs...

The diamond patterns are complete.

Align the twig with the center and mark it so it overlaps the edge by 1/4".

The second twig meets the first in the middle. It needs to be the same thickness and diameter.

15

This is a little high so I'll sand it some more.

on each horizontal member.

When the match is made, nail the twig in place.

Center a short half-round piece vertically on the top member...

Mark the center...

and nail it in place.

Do the same on the other cross members.

I'm going to do the same on each of the other cross members.

Even with these small pieces I always use two brads, never one.

Next I'll put one half-round at the ends of each cross member, flush with the stile.

On the top piece I'm going to add one half-round twig to each side of the center.

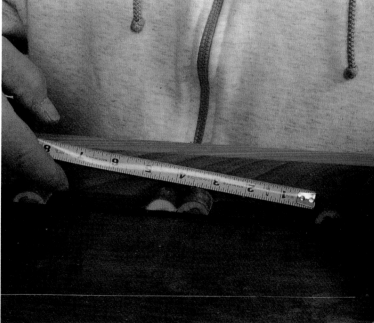

In the spaces between the 3 center half-rounds and the single outside half round, I will center two more. Lay them in place and use a tape for accuracy.

17

When everything is correct, nail them.

The spaces between the half rounds are filled with horizontal pieces.

Continue in the same manner for all the drawer cross members.

If any pieces end up a little high, obstructing the drawer opening, they can carefully be taken down with a sander.

The result.

The legs are cut from 2" diameter stock. They will be about 3" tall.

I bevel the bottom edge of the leg...

Holding it in place, I fire a couple of brads from inside the case to hold it in place.

to give it a finished look.

The legs in place.

To attach the leg, I begin by applying wood glue to the surface.

I then stand the piece up and screw the legs in place from inside the case. Put two screws in each leg for security.

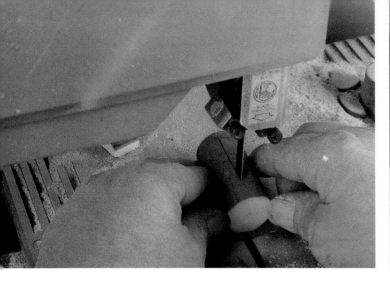

The skirt is made up of half-rounds, the longest being 2-1/2". I split a twig and use one half on each side to assure symmetry.

Repeat on the opposite side.

I also cut four at a time so the second which is angled at the end starts at the same length as the first.

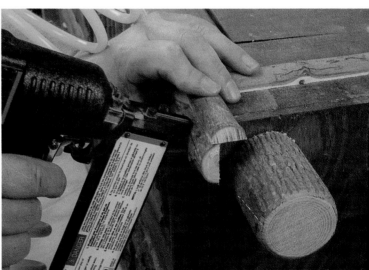

Next set the front corner pieces in place. Put the side one on first and overlap it with the front one. Remember to go from side to side for symmetry.

On the outside twigs I use 3 brads because they are most likely to be knocked off.

From the front corner, I am going to begin my shaping at the third piece on the side....

and the second piece on the front.

On the front skirt, the curve will dip in the middle. After I've put 5 twigs from each corner, I add a twig on either side of the center line, that is the same length as the twigs at the corners.

Continue the curves on the side, making an arc toward the center. Remember to use the two halves of a split twig on opposite sides.

Another long twig with no angle goes on either side of the center twigs.

The finished side skirt.

The third twig from the center has an angle cut on it to begin the arc that will meet that started from the corners.

Continue the curve out five twigs from the center.

Moving back to the front of the chest, I want to continue the line of the horizontal pieces into the stiles. I want the piece to overlap the edge by 1/4".

Complete the curve where the twigs meet.

Make adjustments and nail in place.

The front curve complete.

Progress.

I have created four small panels on the stiles. Picking up the design theme from the sides, I will run a twig along the drawer edge, the height of the mini-panels.

Next add short horizontal pieces at each end of the vertical piece...

As before, by using the other half of the split twig on the opposite side, I maintain symmetry.

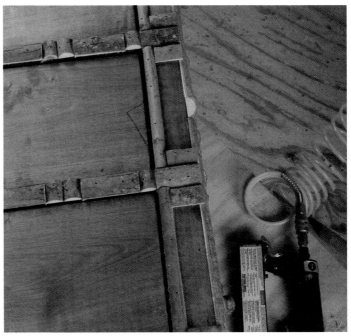

for this result.

Progress.

Measure another twig to fit in the space between the horizontal pieces. Cut, split and sand...

and nail in place.

Another round of horizontal pieces.

Use the other half for the opposite side.

Progress.

Progress.

To avoid a very short horizontal piece, I'll fill the final space with a wider vertical piece.

Adjust the width so it is in alignment with the other pieces.

I am going to use birch bark for the drawer fronts. Often the bark I get appears dark or dirty. I usually prefer to use it in its natural state.

The stile complete.

Progress.

If you want to clean the bark up a little you can rub it lightly with mineral spirits or other solvent. I have even used window cleaner with good success.

When measuring for the birch remember that it will be framed in twig. Here the drawer height is about 7 3/4".

Compensating for the twig I center my measure and determine that a 7" square piece will be adequate.

Always double check before cutting the last side.

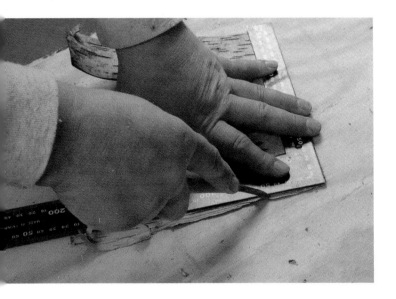

Begin preparing the birch by squaring two adjoining sides.

Cut three squares of birch. Though the square is not my final design, it gives me a uniform place to start.

From these measure and cut the other to sides to size.

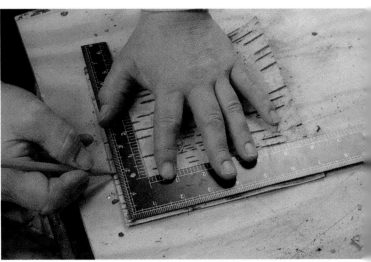

Measure and mark 2" from each side.

Lay an edge between the marks and cut off the corner.

Mark the center of the drawers.

Repeat on each corner...

Cut half rounds the length of the top edge of the birch. Center the birch and the twig at the top...

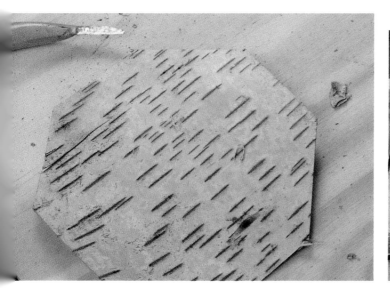

Creating an octagon.

and nail in place.

Align the bottom twig over the lower edge of the birch and nail in place.

Do the same with the vertical sides of the octagons, making sure that they are straight.

Apply the top and bottom twigs to each birch panel.

The ends of the top, bottom and side twigs were cut square. The diagonal pieces are angled to fit.

The result.

Custom fit each diagonal piece.

One birch bark octagon complete.

The octagonal escutcheons are finished.

Continue with the others in the same way.

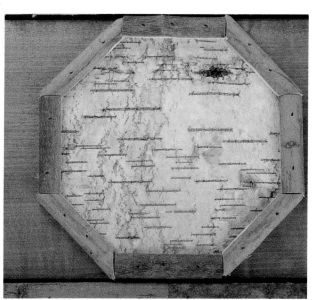

A close-up.

To make the diagonals fit, it is helpful to cut off the points at 90 degrees to the bevel cut.

Fit a vertical piece on the outside edge of the door.

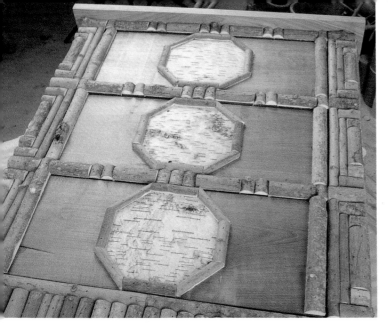

The result.

Cut the end of a twig to the angle of the octagon and fit along the lower edge of the drawer.

and at the top edge of the drawer.

This begins the framing pattern for the drawer front.

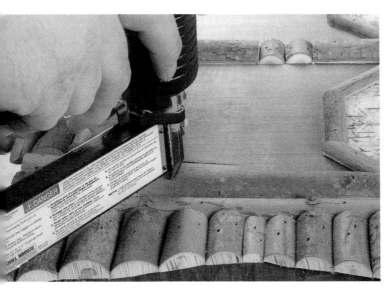

Do the same on the other side of the octagon...

Repeat this step with the other drawers.

It is always easier to set the angle...

Next we add another vertical piece at the outside.

before determining the length.

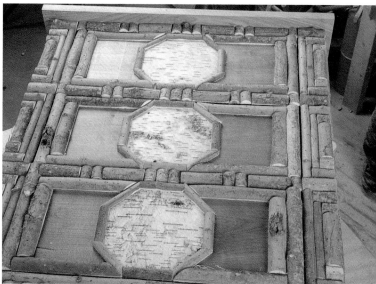

The pattern is emerging.

Progress.

Add the next set of horizontal members, fitting the angled end first.

Add the next vertical piece at the outside.

Progress.

The next set of horizontal pieces is a challenge because of the change in angle at the corners of the octagon.

Another vertical piece...

The solution is to cut the angle on the bandsaw. As before get the correct angle before cutting the length of the twig.

and we're almost there.

Two more horizontal pieces...

leave me with a space for one more horizontal piece.

The last puzzle piece.

Ready for handles.

This unusual imperfection will make a great handle. Unfortunately it is unique, so for balance I will use it on the middle drawer. I mark the cut points on either side, so it will be 6" wide.

I make a groove in the upper surface of the posts....

The other two handles will be plain maple, slightly curved twig, also 6" long.

in which the handle will rest.

The handle posts are 7/8" long.

Apply glue to top of the posts.

Press the handle in place so the curve is up and out slightly.

Mark the center of the posts on the drawer face, for drilling.

Nail the handle to the posts.

Drill through the drawer face.

Position the handles. You may notice that in the process my "up-and-out" handles have been rearranged to go "down-and-out."

Working from the inside of the drawer, countersink the holes.

35

Hold the handle in place and drive in a screw.

With a carving knife or a rotary tool like this, I can make the pattern visible by slightly beveling the outside edge.

Before driving the second screw, double check for alignment.

Do the top and bottom of each element.

One last detail. When I look at the side, there are many potential patterns. I want the focus to be on the diamond patterns in the centers of the top and bottom panels.

This helps bring the eye to the diamond, and will stand out even more with the finish.

Also, I think I'll add a twig edge to the top. Begin by finding and marking the center point in the side.

The front twig should go past the corner...

Nail a short vertical twig to break up the long line.

and be mitered back.

Add a twig from the back to the center.

Measure and mark the center of the front.

Center a small vertical piece.

and check for centering.

Complete the front edge.

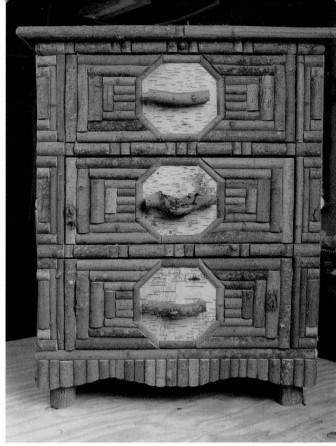

Ready for the finish.

Apply a spar varnish to the top.

Progress.

For the twig work I use satin finish polyurethane. Applied with a large brush, my goal is a generous coat.

Rough places like this handle are particularly challenging.

Work the finish carefully the crevices between the twigs and other uneven surfaces.

When the first coat is thoroughly dried, apply a second.

The Gallery